Animal Cooperation

HALLIE
BLACK *Animal*

Cooperation

A LOOK AT SOCIOBIOLOGY

illustrated with photographs;
diagram by Deborah Roach

William Morrow and Company
New York   1981

Printed in the United States of America.

1   2   3   4   5   6   7   8   9   10

Library of Congress Cataloging in Publication Data

Black, Hallie.
  Animal cooperation, A look at sociobiology.
  Bibliography: p. 60
  Summary: Examines the theory of animal behavior called sociobiology, how it emerged, what it says about animal cooperation, and why its application to human evolution and behavior has raised such controversy.
    1. Animals, Habits and behavior—Juvenile literature.   2. Social behavior in animals—Juvenile literature.   3. Sociobiology—Juvenile literature.   [1. Animals—Habits and behavior.   2. Animal societies.   3. Sociobiology]   I. Roach, Deborah.   II. Title.
QL751.5.B53   591.51   81-1355
ISBN 0-688-00360-5   AACR2
ISBN 0-688-00361-3 (lib. bdg.)

Thanks to Dr. Richard S. Miller, Oastler Professor of Wildlife Ecology, Yale School of Forestry and Environmental Studies, and to Dr. Stan Rachootin, Yale Department of Biology, for reading over the manuscript of this book.

Credit for Photographs

Charles H. W. Foster, 39. James Moore, 42. National Oceanic and Atmospheric Administration, 43 top; William L. High, 43 bottom. Mark Plotkin, 38 bottom. United States Department of the Interior Fish and Wildlife Service, E. P. Haddon, 44 top; E. R. Kalmbach, 37 top, 44 bottom; Bureau of Sport Fisheries and Wildlife, L. C. Whitehead, 37 bottom. Rick Weyerhaeuser, 38 top, 40, 41.

To Hank

Contents

1 Sociobiology 9
2 The Roots of the Theory 13
3 Animal Behavior 22
4 Another Side to the Story 34
5 Human Behavior 48
6 The Final Word 57
  Bibliography 60
  Index 63

1

Sociobiology

Animals can seem almost human sometimes. Elephants twine their trunks as if kissing. Chimpanzees tickle each other in play. Dolphins rescue an injured companion by pushing it to the surface so it can breathe.

Animals can seem generous, cooperative, and affectionate. Why do they behave this way? How is our behavior different from theirs? Scientists have long puzzled over these questions. Now a new field of study, sociobiology, is trying to answer them.

Most sociobiologists have been trained as biologists, but they study more than animals' biologic functions such as eating, sleeping, and mating. They study animals' social life, how animals form groups and cooperate. This special interest gives the new science its name.

Sociobiologists have a particular point of view about animal social life. Whether they study ants or elephants, they believe that everything an animal does in its social life serves a purpose for survival. Otherwise natural selection would have eliminated the behavior long ago. They believe, too, that patterns of social life are inherited by each animal species just as is eye color or body shape.

It is not always easy to find a purpose for what animals do, though. Sociobiologists try to understand animal behavior through a kind of scientific detective work. They gather information about an animal's habitat, how the animal finds food, and what dangers the animal faces, such as predators. This information comes from ecologists who study a particular area or from biologists who are observing a particular animal.

Using this information, sociobiologists try to explain why, for example, chimpanzees tickle each other. Observers have seen one chimpanzee tickle another until the second chimpanzee collapses and rolls on the ground, its shoulders heaving in silent laughter. (Chimpanzees cannot laugh aloud as human beings can.) One explanation is that tickling helps individuals in the troop recognize each other. Another explanation is that tickling promotes group cooperation.

Chimpanzees live in dense forests in troops that can number up to eighty individuals. Often they move about in small groups, however. Only at intervals does the whole troop assemble. Nevertheless, each chimpanzee knows the members of its own troop. It can tell if an approaching animal is a stranger or not. Tickling bouts are one way it learns to make that distinction. In tickling bouts, young chimpanzees come close together. They touch each other and become familiar with each other's features and voice. Eventually they learn to recognize troop members at sight.

But tickling is not just something youngsters do in play. Mothers tickle crying, hungry, or fretful infants to distract them. One adult tickles another who is angry or sad. Tickling bouts are more than a way to have fun; they are a form of communication and a way to show concern. They help cement the social life of the troop.

Sociobiologists are opening a new perspective on animal social life with such analyses. But they are also arousing controversy. They are doing so by applying their theories to human social life as well.

Many people, scientists included, consider human beings unique. Sociobiologists do not. They believe that human behavior is very similar to that of other animals. They see human social life as a set of patterns inherited from mammal ancestors, especially monkeys and apes. For example, some think human females have an inherited tendency to be dominated by males, which is a kind of social pattern observed among baboons.

Such a view of human behavior has aroused strong opposition to sociobiology. Many scientists think human behavior is not inherited but is shaped by the experiences one has growing up. This belief is called the "nurture" view. It places great importance on what a person learns from parents, relatives, friends, and people in the community. Thus, according to the nurture view, girls are not naturally passive, but they may become passive if they are raised to behave that way.

Sociobiologists, on the other hand, think that inherited nature determines behavior just the way it determines many physical features.

The nature versus nurture debate is an important one to scientists. It is part of their effort to understand how the natural world evolved and what place human beings occupy in it. Are we unique or are we "naked apes"? Sociobiology is one of many attempts to solve this riddle. The story of sociobiology illustrates how scientists develop and test new theories.

The debate is also very important outside the world of science. Its social impact is as great as the discovery of a new medicine or a new source of energy. It can affect the view of social problems such as poverty and crime. If people think these problems result from a child's poor upbringing, or nurture, they may try to help the parents or community do better. They may propose special programs to provide jobs, education, and other help. They may try to rehabilitate criminals.

However, if people believe in the nature view, they assume that some people are born lazy or dishonest. They may then consider attempts to change them or to eliminate poverty and crime impossible.

Because sociobiology can affect how society views human nature, it has aroused a great deal of concern among scientists and nonscientists alike. Many people consider it a dangerous explanation of behavior, which could appeal to people with racist or sexist prejudice.

You will be hearing about sociobiology in the future. The following chapters explain how it emerged, what it says about animal cooperation, and how it affects our understanding of evolution and human behavior.

2

The Roots of the Theory

Sociobiology is a new word coined in the early 1970's. But the roots of the science go back over one hundred years to the theory of evolution worked out by Charles Darwin.

In 1859, Darwin wrote *The Origin of the Species*, in which he described his theory of "descent with modification." He saw nature as the scene of an immense struggle for survival. Individual animals vie for food, water, shelter, mates, and other necessities of life. Animals that can't compete well will be weeded out. This weeding-out process Darwin called "natural selection."

Although people in Darwin's day knew little about heredity, they were aware that parents pass on certain traits to offspring. Therefore, Darwin reasoned, the traits carried by the poor competitors would gradually be eliminated since those animals would not survive to produce offspring. The traits carried by the survivors would be passed on and accentuated over time. In time, over many generations, this modification would lead to the astonishing variety of plant and animal species that exists in the world.

Darwin's theory emphasized the role of competition in animal life, but there is striking evidence that animals cooperate, that they even risk their lives for one another.

A bird cries out when it sees a cat. Prairie dogs chatter and bark when they sight a hawk. When a pronghorn antelope senses danger, it jumps in such a way that its white rump flashes. This behavior sends the rest of the herd running. In each case, the animal endangers itself to warn the rest of trouble.

Other ways animals cooperate are less dramatic but just as important. Hyenas, elephants, and chimpanzees will adopt orphaned young. Lionesses will nurse each other's cubs instead of saving their milk for their own. In each case, an adult takes time and attention from its own offspring.

Such generosity, which scientists call "altruism," puts an animal at a disadvantage in the competition for survival. It is not devoting full time and energy to competing. It may even be helping potential rivals. If animal behavior is coded by genes, then those genes—"generous genes" in scientists' slang—should be eliminated by natural selection. The generous animal should lose out in the competition, and its generous genes should perish with it. Yet altruism appears in generation after generation.

To most scientists, the appearance of altruism from generation to generation was proof that there were no such things as generous genes. Therefore, altruism had to be something that animals learn. Then, in 1964, a biologist named William Hamilton published a theory as to how generous genes could be perpetuated. He reasoned that many individuals in a group are related to each other. Thus, they share a common genetic heritage through grandparents, aunts, uncles, and so forth. A generous act that endangers one bird or antelope will save some of its relatives, which will pass on generous genes to the next generation.

Hamilton's theory gave biologists a new way of looking at animal cooperation. By 1971, he and others such as J. Maynard Smith and Robert Trivers had worked out a new version of Darwin's theory of natural selection. According to this new version, selection not only weeds out poor competitors, it weeds out families whose members don't cooperate.

Darwin had seen survival as a matter of competition. His supporters described evolution as a game of "one-on-one." The new theory pictured survival as the result of team play. The team is made up of relatives. They cooperate to protect each other from danger and to acquire food, shelter, and other necessities. The better their cooperation, the better their chance of survival.

Hamilton called this new version of natural selection "kin selection." He saw kin selection as a force for the evolution of social life among animals. He and other scientists who agreed with him became known as sociobiologists.

Sociobiologists looked for proof that kinship is connected to cooperation. Some interesting evidence came from observations of wild turkey flocks in Texas. In these flocks brothers apparently helped each other attract mates.

This teamwork begins when the brothers are six or seven months old. First they wrestle with each other. Such behavior may seem an unlikely source of cooperation, but it enables the brothers to establish which one is stronger. The distinction becomes important later.

The next spring, when the brothers are sexually mature, they strut and display their feathers together as the females gather and watch. Other pairs of brothers do the same. If a female is attracted to a pair, the weaker brother steps aside. Only the stronger brother mates.

The weaker brother voluntarily gives up a chance to mate and produce offspring. However, by helping his stronger brother attract a mate, he increases his chance of having nieces, nephews, and cousins. Sociobiologists do not know how or why this kind of teamwork evolved, but single males seem to have little chance of attracting a mate. The stronger brother needs the help of the weaker one.

This evidence that kinship plays a part in animal social life has encouraged sociobiologists to study many mysteries of animal behavior. One mystery they have tackled is why bees, wasps, and ants form such cooperative groups.

Bees, wasps, and ants are known as social insects because they form miniature societies. Each member has a certain job, such as defending the nest or gathering food. In addition, only one female, the queen, produces offspring. The other females care for the queen and tend her young rather than produce their own. Darwin called social insects "the one special difficulty, which at first appeared to me actually fatal to my theory." He was not sure how these insects avoided competition.

Now sociobiologists offer an explanation. The answer, they suggest, lies in the unique genetic closeness of female wasps, bees, and ants. They have an unusually close kinship. Most of them are sisters, but they are more like identical twins than normal sisters.

Identical twins are formed when an egg splits after it has been fertilized by a sperm cell. Two individuals develop instead of one. The reason such twins look alike is that they share one hundred percent of their genes. They have identical genetic codes for eye color, hair color, and all other features that are inherited through parents.

Normal sisters, by contrast, share only a portion of their genes. Each child gets a different assortment of genes from its parents. Genes exist in pairs, but parents do not transmit whole pairs to their children. Instead, each parent transmits only one gene from each pair. This assortment is carried by the parent's egg or sperm cell. When the egg is fertilized, the genes from each parent match up to form full pairs again.

Although the assortment received by each child in a family is different, children of the same parents are likely, on the average, to possess half their genes in common. They share half of all the genes they receive from their mother—one quarter of their total—and half of all the genes they receive from their father—a second quarter of their total. The two quarters add up to half of all their genes.

Bee, wasp, and ant sisters fall halfway between identical twins and normal sisters. Like human sisters, each receives a different selection of genes from its mother. Like human sisters, each shares one quarter of those genes with her sisters. But all the sisters receive identical genes from their father. This pattern is the result of an unusual feature of reproduction among such insects. The males develop from unfertilized eggs. They don't have full gene pairs but only one gene for each. Thus, the males cannot transmit a different assortment to each daughter. Their sperm cells are all identical.

As a result, these insect sisters share all the genes they receive from their father, one gene of every pair they have. In addition, they share some of the genes they receive from their

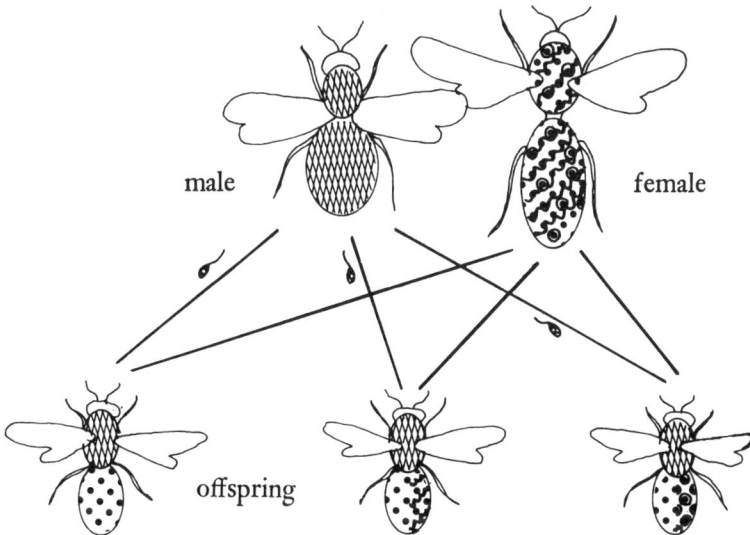

The male bee gives his daughters identical genes.

mother. In total, they share three quarters of all their genes. On their mother's side, they are normal sisters, but on their father's side they are identical twins.

Sociobiologists do not know how this unusual system of reproduction evolved, but they do consider it an important clue to the cooperation of social insects. Sociobiologists reason that insects with this system of reproduction have a very close kinship, closer than the kinship of ordinary sisters. In addition, females make up about 90 percent of a hive or nest and are all descendants of the queen or her relatives. A wasps' nest or beehive is like a community of nearly identical twins. Sociobiologists think that this closeness enables the females to cooperate with each other rather than compete.

Of course, sociobiologists know that among some social insects cooperation isn't all voluntary. The honeybee queen, for example, produces a chemical substance, a hormone, that prevents the other females from reproducing. However, sociobiologists see this hormone as a kind of backup that evolved later. They think species that lack such backups are more primitive forms of social insects. These primitive forms live in social groups that are often less permanent and less cooperative.

The paper wasp is an example. The queen starts a nest each spring with the help of a few sisters. They feed her, help build a nest, and tend her brood. But they also try to lay their own eggs. If the queen finds these eggs, she eats them. Nevertheless, the sisters stay with her through the summer. Then, in the late summer, they leave her to mate on their own. The group dissolves for the year and must be reestablished again next spring. Honeybee societies, by contrast, remain year after year.

Do these insect sisters know they are related? Only recently have scientists designed experiments to determine whether they do or not. In 1979, Les Greenberg, an entomologist at the University of Kansas, reported the results of one such experiment. He decided to find out if bees use scent to tell relatives from nonrelatives. Social insects have a highly developed sense of smell. Studies have already shown that hivemates know each other by a common odor. Guard bees, wasps, or ants patrol the entrance to their hive or nest and turn away individuals that don't have the proper smell.

Greenberg thinks hivemates may inherit this common odor. He demonstrated the possibility by taking young bees from the individual cells in which they had developed before they emerged and met adults. He raised the young bees separately and bred them into new colonies. Then he took their descendants back to the original hive. The guards, which were their cousins or aunts, let many of them in even though they were strangers. Greenberg thinks the bees raised separately were still recognized as relatives by their smell.

Sociobiologists have turned to other aspects of animal cooperation. Scientists have long noted that the males and females of most species do not share the task of raising the young. One parent usually does the job alone. Why? Such a lack of cooperation seems dangerous. It deprives the young of the care and protection that a second adult could provide.

Sociobiologists interpret the behavior differently. According to them, each sex has a different strategy for increasing its descendants. That strategy determines which sex cares for the offspring and whether both parents will cooperate.

The female often carries the fertilized egg or the developing young in her body. She is the one that deposits the eggs or gives birth to live young. Sociobiologists reason that caring for the young after that point to increase their chance of survival is to her advantage also. That way she increases her prospect of having children, grandchildren, great grandchildren, and so forth.

But the male needs a different strategy. He only fertilizes the egg. He best increases the number of his descendants by mating with as many females as possible. He is also at a disadvantage compared to the female. She can tell which off-spring are hers, but he often cannot. According to socio-biologists, helping to raise the young would be risky for the male. He might actually help care for the offspring of a rival.

An exception that seems to prove the rule is the three-spined stickleback, a tiny fish studied by biologist Niko Tinbergen. Among sticklebacks, the male, not the female, is the one that cares for the young. This arrangement makes sense to socio-biologists because the male, not the female, is the one that can tell his own offspring.

The males establish individual territories and defend them fiercely against other males. During breeding time, a male builds a tunnel in the sand. He lures several females into it, one at a time. There each female deposits eggs, which he fertilizes. When the eggs hatch, the male accepts them as his. After all, no other male has been near the tunnel. But a female can't tell which are hers and which belong to other females who entered the tunnel. So she doesn't help care for them. The male is the one that blows water into the tunnel to remove wastes and to force in tiny organisms for the young fish to eat.

An unusual variation in the way offspring are reared is seen among birds like cuckoos and cowbirds. They are parasites that lay their eggs in the nests of other species. They leave the job of raising their young to strangers. What has long puzzled scientists is the response of the foster parents. Surprisingly, some don't prevent the parasite from laying her egg in their nest. They raise the young parasite even though it takes food and attention from their own young.

These foster birds may appear to be bad parents to their own young, but sociobiologists interpret their behavior as a form of kin selection. Evidence comes from the study of one parasite/foster bird relationship by biologist Neal G. Smith.

In 1968, Smith reported a study of chestnut-headed oropen-dula, a bird parasitized by cowbirds. Oropendula are natives of Central America, where they form colonies in very high trees. From this vantage point, they can easily spot approaching cowbirds. Yet the oropendula seem to let cowbird females enter boldly and lay their eggs.

When Smith observed what happens in the nest, he made a startling discovery. He found that oropendula young actually benefit from the presence of cowbird young. Apparently, many oropendula young die of botfly infestations. The adult botfly lays eggs on the chick's skin. When the eggs hatch, the botfly larvae burrow into the chick's body and kill it. But cowbird chicks have the ability to spot and pick off the larvae from their nestmates. Thus they reduce the danger of infestation. True, some oropendula chicks die of starvation because the young cowbirds deprive them of food. But more would die of botfly infestation if there were no cowbirds in the nest.

Sociobiologists believe that oropendula adults are sacrificing some of their young in order to increase the survival chances of the rest. In this way, sociobiologists explain the failure of the adults to guard their nests against cowbirds. The oropendula adults are engaging in a kind of kin selection.

Such explanations of puzzling animal behavior are attracting scientists from many fields of biology to sociobiology's views of animal cooperation. The best known sociobiologist today is Edward O. Wilson, a Harvard professor who was originally trained as an entomologist and who is an expert on social insects.

He has written two books that greatly expand the application of sociobiology. In the first book, *Sociobiology*, published in 1975, he speculated that sociobiology can explain all aspects of social life, animal as well as human. In the second book, *On Human Nature*, published in 1978, he speculated on how sociobiology can explain human culture. These books moved sociobiology far beyond the world of honeybees and wasps. They also set the stage for recent controversy.

3

Animal Behavior

How do animals cooperate? One important way, according to Wilson and other sociobiologists, is to form groups for mutual protection and for help in finding food and caring for the young.

This grouping may come about accidentally. Here is William Hamilton's description of a situation in which animals might form temporary groups without meaning to: "Imagine a circular lily pond. Imagine that the pond shelters a colony of frogs and a water snake. The snake preys on the frogs but only does so at a certain time of day—up to this time it sleeps on the bottom of the pond. Shortly before the snake is due to wake up, all the frogs climb out of the water. This is because the snake prefers to catch frogs in the water. If it can't find any, however, it rears its head out of the water . . . and snatches the nearest one."

Hamilton explains that "a frog that happens to have climbed into a wide-open space will want to improve his position." So will all the other frogs. The result will be a jumble of frogs each jockeying for a protected inner position. Frogs in the bottom of the jumble will be safe from the snake. Frogs that hover at the outer edge or that stand alone will be caught.

Hamilton suggests that the trait of pressing into the jumble, or not, is genetically coded. The frog that stands alone will be killed, and its genes for that kind of behavior eliminated. In time, the only surviving frogs will be ones that carry a genetic code for crowding together. Hamilton thinks a process like this one could cause a herding instinct to evolve among some animals.

But Hamilton's lily pond is imaginary. Sociobiologists must now translate his ideas into facts. Animals form an enormous variety of groups—packs, prides, flocks, and so forth. Sociobiologists must explain why this variation occurs.

Sociobiologists start with the assumption that every aspect of an animal's social life has a purpose. Thus, they believe that each kind of group evolved to help each species adapt to the particular conditions of its habitat and way of life. A primary purpose of the social group, they suspect, is to help each sex within the species adjust its strategy for kin selection.

To prove this theory, sociobiologists are reviewing what is known about particular animals in order to identify the reasons an animal forms the kind of group it does. To make such a review possible, sociobiologists often use research carried out by scientists who may not be sociobiologists. Many important animal studies were begun or completed before sociobiology emerged. Others are being carried out by scientists who do not accept sociobiology. Nonetheless, sociobiologists are reinterpreting these studies in the light of their theories about animal behavior. Here are some examples of their reinterpretations.

Vicuña

The vicuña is a graceful animal of the high Andes Mountains in South America, somewhat similar to a llama. Its social life, described by biologists Carl B. Koford and William L. Franklin, fits sociobiological theories.

Vicuña form family groups of one male, several females, and their young. Each male has a territory that he defends against other males. He keeps his family inside his territory. The territory is dotted with mounds of dung. As part of their daily activity, members of the family visit the mounds, sniff them, and add to them. Outside the territory are groups of twenty or so bachelor males. They were driven from their families when they became sexually mature and have not yet acquired their own territories.

Sociobiologists explain that this social behavior can serve a purpose. They point out that vicuña are grazing animals. Their habitat is a fragile environment with thin soils and scarce grass. This habitat could easily be destroyed by overgrazing. Sociobiologists think vicuña family groupings prevent this destruction. Each family stays within its own territory. The male even moves his family from spot to spot within his territory. Such movement gives the land a chance to recover from grazing and to produce new grass.

Thus, vicuña social life is like a rationing system. Each family has a certain area to use. This analysis may explain why the dung piles are so important. They are markers that enable members of the family to identify their own territory. And if the markers are not enough, the male is ever on patrol, keeping his family in and strange males out. If he didn't patrol his boundaries, strange males might slip in and mate with the females. Then the male would be providing grazing space to the offspring of rivals.

African Wild Dogs

African wild dogs are small, skillful predators. They do not really look like domestic dogs, which are descended from wolves, but they do belong to the same family, Canidae. They are rare, secretive animals, but several biologists have recently been able to observe their social lives. Among these biologists are Richard Estes and the Baron Hugo van Lawick.

According to their descriptions, African wild dogs form highly cooperative groups. They live in packs of up to forty animals. Males and females pair up to mate, and the group is closely related. Wild dogs do not form rigid family groups, though. All the members of the pack mix together. The pack is like a family clan.

Most of the time members of the pack are very cooperative. They feed companions that are too ill or old to hunt. When prey is brought down, adults let the pups eat first. Adults can also regurgitate partly digested food if need be. If a hungry pup begs for food, an adult can provide some by regurgitating. The pack is like a communal day-care center, with each member looking out for the others.

Sociobiologists trace this kind of social life to the way African wild dogs provide food. They specialize in hunting animals like gazelles, which are very swift, or like wildebeest, which are very large. This strategy requires a high level of teamwork, and it also requires the services of many members of the pack. The dogs must be numerous enough to run down and surround their prey. They must be able to coordinate their actions during the hunt. As it is, teamwork is carried to such a degree that even nursing mothers can help hunt. While they are away, another nursing mother will feed their pups.

However, the family life of wild dogs is not completely harmonious. Van Lawick reported seeing one mother systematically kill the pups of another. This observation seems to contradict the image of the pack as one big happy family, but sociobiologists have an explanation.

They point out that the first mother's action is a form of kin selection. She is promoting her own offspring's survival at the expense of another family. By killing the other pups, she is removing competition for food and care. Once the pups are dead, the adults of the pack will concentrate their attention on her young.

Observations like van Lawick's lead sociobiologists to conclude that the most harmonious societies are those made up of close relatives, like the beehives. Once strangers or more distant relatives join, the door is open to greed, jealousy, and spite. These emotions, which we consider negative, are seen by sociobiologists as inevitable because they play a useful part in kin selection among rival groups. Such feelings enable members of one family to carry out actions that harm other families. They are the reverse side of the coin of generosity and cooperation.

Lions

Most cats are solitary animals and hunt on their own. Lions are an exception. They live and hunt in groups called "prides." Why are they different? Sociobiologists find a clue in their way of life. Lions hunt large animals, such as wildebeest, which one lion alone can rarely bring down. Like wild dogs, lions must rely on teamwork.

Lion family life has been studied by many biologists, among them Richard Estes and George Schaller. From their descriptions, lion family groups are almost as cooperative as wild dog packs, but there are some important differences.

A pride can contain from fifteen to thirty animals on the average. However, the females outnumber the males, sometimes four to one. The females do the hunting and care for each other's cubs. Usually one female "baby-sits" for all the cubs while the other lionesses hunt. Lionesses, which will even nurse each other's cubs, cooperate to a high degree.

But the males, by contrast, seem downright selfish. They rarely help hunt. And while they sometimes play with the cubs, they don't participate in raising them. In fact, when prey is caught, the males push and shove to eat first. Cubs are sometimes trampled to death in the crush. At times, cubs get so little to eat that they die of malnourishment. More startling, Schaller reported that a young male defeated the old leader of the pride and then killed all the cubs.

Sociobiologists have tried to explain why females should be so cooperative but males so selfish. They think one answer lies in the pride's mixture of relatives and strangers. The core of the pride is a group of related females—aunts, daughters, sisters, cousins, and so forth. According to sociobiologists, this genetic closeness accounts for their cooperation.

Males, on the other hand, are usually newcomers. Young males leave the pride in which they were born when they are sexually mature. Their place is taken by other males who have also left their original pride. The newly arrived male has no kin ties to the existing cubs. According to the theory of kin selection, removing those cubs to make room for his future offspring is to his advantage. And since he lacks kin ties to the other adults, he has no inclination to exert himself for the sake of the pride. He lets the females hunt and shoves in only when they have brought down their prey.

Elephants

Like vicuña, elephants are vegetarian. They browse on leaves. Therefore, they do not need to cooperate to hunt. And their size protects all but the youngest from predators. A full-grown African elephant is the largest land mammal alive today. Seemingly, elephants could survive nicely on their own. Yet they form highly cooperative groups that are so stable that they are like family dynasties.

Looking for a reason, sociobiologists point to the long period during which young elephants mature. A newborn elephant requires a great deal of care. Although it can walk soon after birth, it must learn the proper way to use its trunk for gathering food and drinking. Its tender skin would be burned by the sun if adults didn't spray it periodically with dust and water. As young elephants grow up, they must also learn where the best feeding grounds are and what routes to take from one to another.

The care and teaching a young elephant needs is provided by the herd. Sociobiologists think it is no coincidence that all the adults in the herd are female. This fact fits with their belief that females have the highest stake in caring for the young. Male elephants, by contrast, leave the herd when they become sexually mature and either live alone or join other bachelors. Although the males rejoin the females during mating time, the female herds keep separate the rest of the year.

Sociobiologists also find it significant that the females in each herd are all related. They are cousins, sisters, aunts, and so forth. Their leader is the oldest female. Usually she is also the biggest, since elephants continue to grow after sexual maturity. An elephant herd is thus a close kinship group, which serves the purpose of rearing the next generation.

As described by biologists such as Iain Douglas-Hamilton, elephant social life is harmonious and affectionate. Mothers help feed and look after each other's calves. In addition, young females that have not yet mated become "aunts" to the calves of others. When the herd moves from place to place, the young travel in the middle of the procession. If danger threatens, the adults quickly tighten the protective ring and the leader prepares to charge.

Elephants also care for members of the herd that are sick or very old. When an elephant is too ill to stand, others try to prop it up. If it lay down, its own weight would crush its lungs. Most remarkable, when an elephant dies, other elephants gather round and push the body out of sight. Perhaps this behavior is what has given rise to myths of a secret elephant burial ground covered with ivory tusks.

Dolphins

Comparing dolphins to vicuña and elephants may seem far-fetched, but sociobiologists see similarities. Like elephants, dolphins form cooperative herds. Adult females care for each other's young, also called "calves." Adolescent females become "aunts." Dolphins, too, look after sick or injured companions. Air-breathing mammals, they must surface periodically. A dolphin that can't will be pushed to the surface and held there so that it has a chance to recover.

But like vicuña, dolphins form family groups with males. However, these groups are not permanent and the males do not have territories. The females travel with their young, and the males join them for mating. The males also stay with their families when the calves are young. Bachelor males form their own groups.

The resemblance of dolphin social behavior to that of vicuña and elephants may stem from a common way of life, sociobiologists propose. All three live by grazing, or browsing, if you think of dolphins as "grazers" of the sea. They roam from spot to spot in search of large schools of fish to feed on. Since they are prey animals themselves—killer whales, for one, hunt them—they form large groups of up to one hundred animals and find safety in numbers.

Baboons

The first description of baboon social life came from observation of caged or confined animals. Observer after observer reported rigid, often violent behavior. Males fight for dominance. Once a leader is established, he tries to gather the eligible females into a harem and exclude all the other males. Often all the males join in to bully the females, which seem to have no independence at all.

Now studies are being done on baboons in the wild. Although the earlier generalizations seem to hold for one baboon species, the hamadryas of Ethiopia, observers have found that baboon social life varies from species to species. Furthermore, even hamadryas behavior is not as simple in the wild as it is in captivity.

Studies of the hamadryas show that a male in the wild forms a family with several females and their young. Within the troop, males do fight for dominance. They also spend a great deal of time keeping the females of their family together. If a female tries to break away, the male chases her and bites her until she returns submissively to her group. Despite this preoccupation with females, bachelor males are tolerated even when they become sexually mature.

Sociobiologists trace the baboon's social behavior to its way of life. Baboons are mostly vegetarian. Many species live in grasslands or savannahs. There they must travel long distances in open country to find food. They are in danger from predators, yet they can't count on climbing trees for escape. Thus, according to sociobiologists, baboons must defend themselves. However, self-defense is easier for the males than the females as the females are burdened with infants and young offspring.

Sociobiologists think the rigid family grouping and strong male domination have evolved to protect baboons, especially females and young, from danger. The domineering males keep the females together in a tight group so one doesn't stray into danger. The bachelor males are tolerated because they give the group additional protection.

Sociobiologists speculate further that the females have evolved into passive and submissive animals as such behavior makes them easier to keep together in a group. The males, however, have become more aggressive and dominant to fit them better for fighting. They even have sharp incisor teeth, which the females lack, to use as weapons.

Chimpanzees

Chimpanzees, like baboons, were first studied in captivity. There they seemed to form family groups of one male and several females and to spend much time establishing dominance ranks. But studies done of wild troops now present a much different picture. Biologists such as Jane Goodall, who has studied a population of chimpanzees in the Gombe Stream Reserve of Tanzania for over twenty years, report a flexible kind of social life.

Chimpanzees form large troops of thirty to eighty animals. But the troop rarely travels as a unit. It breaks into small groups. Adult males travel together; so do mothers and offspring. The membership of the groups shifts frequently. Sometimes a grown male will rejoin his mother and younger brothers and sisters. Sometimes males will be joined by females.

Troop life is highly cooperative. For example, groups spread out in search of food. When one group finds a fruit-laden tree or some other delicacy, it hoots for the rest of the troop. As the other groups arrive, the chimpanzees nuzzle and hug each other. Then they share the feast.

Chimpanzee social life seems relaxed and affectionate. Chimpanzees spend a good deal of time napping in trees, resting on the ground, playing, tickling, and grooming each other. Although the males struggle to establish dominance ranks, the leaders don't direct the life of the troop as baboon leaders do.

Sociobiologists think this flexibility results from the kind of habitat chimpanzees occupy. They live in dense forests where food is abundant. There is plenty of natural cover to hide in. There are many trees to climb. So, sociobiologists speculate, chimpanzees can spread out in safety. They do not need to form tight groups or maintain a constant guard.

An interesting fact is that wild chimpanzee males do not form families. A few times a year mature females are receptive to mating. The timing of this period, called "estrus," is controlled by each female's hormone cycles. During estrus, a female mates with many males. The males cannot distinguish their own offspring. Nevertheless, adult males take great interest in the young. They play with them, break up games that get too rough, and protect them in danger.

This behavior seems to contradict sociobiological theories. The males ought to ignore the young and leave care to the females. Why are they affectionate? One explanation sociobiologists offer is that the troop is a large kinship group. Although an adult male may not be able to distinguish his direct offspring, he may recognize the young as kin—cousins, nieces, and so forth.

The importance of kinship shows up in many ways in chimpanzee social life. Brothers and sisters remain very close even as adults. Brothers form alliances to help one another achieve dominance. Sisters stay together when they have their own offspring so cousins often grow up together. And even as adults, chimpanzees remain close to their mother. When a new infant is born, older brothers and sisters act as baby-sitters. If the mother dies, an older brother or sister adopts the family. Goodall even saw an adolescent male try to raise an infant brother.

To sociobiologists, analyses and reinterpretations of animal social life, such as these examples, open an exciting new prospect. They hope to draw up categories of animals according to the kinds of social groups they form. In a similar effort over 200 years ago, scientists sorted out animals according to physical traits, classifying them into orders, families, genera, and species.

The new categories would cut across these old divisions. Dolphins would be compared to other grazers, lions to other animals that hunt in groups, and so forth. Such categories would give scientists a new way to think about animals and how evolution shaped their behavior.

4

Another Side to the Story

The prospect of new categories of animals does not please every scientist. There are those who are not convinced of sociobiology's claims. One reason is that there exists a great difference between studying an animal's physical traits and studying its behavior. Each scientist can examine an animal's circulatory or reproductive system and agree with other scientists on its anatomy. But agreement on a description or interpretation of an animal's behavior—especially when one scientist must take another's word for what the animal did—is much harder.

This difference is especially important for sociobiology, since the discipline rests its case on interpretations. For example, it is not a fact that generous genes exist. No such genes have ever been found. Rather, the idea of generous genes is based on interpretations of animal behavior that others see differently.

Thus, what sociobiologists see as altruism, other scientists see as selfishness and competition. Take the bird that gives warning, for example. Critics of sociobiology doubt that the bird is really trying to warn companions. Instead, it may just be looking out for itself. By alerting the others to fly away, it creates

34

a diversion. The commotion may confuse the cat and give the bird a chance to slip away in the crowd. As for young females who baby-sit, they may not be trying to help the mothers. They may just be trying to gain experience for raising their own off-spring.

A second criticism is that sociobiologists have oversimplified their descriptions of animal behavior so that all the evidence fits their theories. Contradictory facts are said to have been over-looked.

For example, much observed lion behavior does not fit the pattern of cooperation outlined by sociobiologists. For one thing, females as well as males can change prides. For another, females compete as fiercely as the males to eat first. In fact, mothers have been seen pushing their own cubs away from the kill. George Schaller thinks that the selfishness of the males may actually benefit the cubs at feeding time. By driving off the females, the males—who are fewer in number—make room around the carcass for the cubs.

Observers of baboon social life also present evidence that contradicts what sociobiologists emphasize. They have found that the females, not the males, are really the center of power in the troop. Males may fight for dominance, but the leader is really chosen by the females. This pattern was not easy to see at first. The fighting of the males is obvious. The way the females select a leader is subtle. They pick out a male who shows concern for their young.

Observers even question how helpful male baboons are in protecting the females and young. One report indicated that the first reaction of baboons to danger was flight. The males raced away, leaving the females to lumber along as best they could with infants and older offspring in tow.

A third criticism is that animal behavior may not be geneti-cally coded as sociobiologists believe. There is evidence that animal social life is flexible and changeable.

35

For example, University of Colorado researchers Marc Beckoff and Michael Wells have reported this finding for coyotes. For three years, 1977 to 1979, they observed coyotes near Blacktail Butte in Wyoming's Grand Teton National Park. They observed that coyotes sometimes live in packs, sometimes in pairs, and sometimes alone. What's more, the same coyote may live all three ways in its lifetime.

The kind of social group coyotes form seems to depend on the supply of their favorite food, which is rodents. When animals such as field mice and ground squirrels are abundant, usually from May to October, coyotes live alone or in pairs. But the rest of the year, when rodents are scarce, they gather into packs.

During that period of scarcity, from November to April, coyotes must live on whatever dead animals or live prey they can find. Beckoff and Wells think that the cooperation of a pack enables them to find food more easily. By grouping into a pack, several coyotes are able to stake out a territory and try to establish exclusive right to live prey or carcasses within it. Animals in the pack share the tasks of patrolling the boundaries and searching for food. They bring food to pregnant females or mothers with nursing pups.

In the spring, when rodents become available again, the packs break up. Interestingly, though, coyote males and females remain in pairs, and the members of the pack may all be offspring that have stayed in the area.

More evidence of changeable behavior comes from Thelma Rowell, a biologist who has spent over twenty years studying baboons. She has observed many species besides the hamadryas. She found that each troop has its own social pattern, regardless of similarities of habitat and way of life. Some species living in grassland or savannah formed rigid groups; others didn't.

She also found that social life, especially the behavior of each sex, changed when animals were caged. She tells of a hamadryas

One way animals can cooperate is to protect each other. Prairie dogs live in "towns" that are exposed to danger from coyotes, hawks, and other predators. But prairie dogs help each other by keeping watch and making a lot of noise when they spot danger. Only when they have warned other prairie dogs of a threat will they dart into their own burrows for safety.

Above: The African wild dog is a skillful predator that can bring down very large animals, yet it is a small animal itself. The key to its success seems to be teamwork with other members of the pack.

Right: Vicuña live in a harsh environment with little grass to eat. Sociobiologists think vicuña males guard specific territories in order to provide their families with private grazing areas.

38

The core of a lion pride is a group of related females, which share the work of bringing down prey and looking after cubs. The lucky cub in the lower picture has part of a carcass all to itself. Normally, cubs have to wait their turn until the adults have eaten, and many cubs die of malnourishment.

An elephant herd is made up of related females who share the work of rearing the young. The oldest female is the leader. When she moves the herd from place to place, the young elephants always travel in the midst of the adults for protection. When the herd stops at a watering hole, older females help the young ones learn to drink with their trunks.

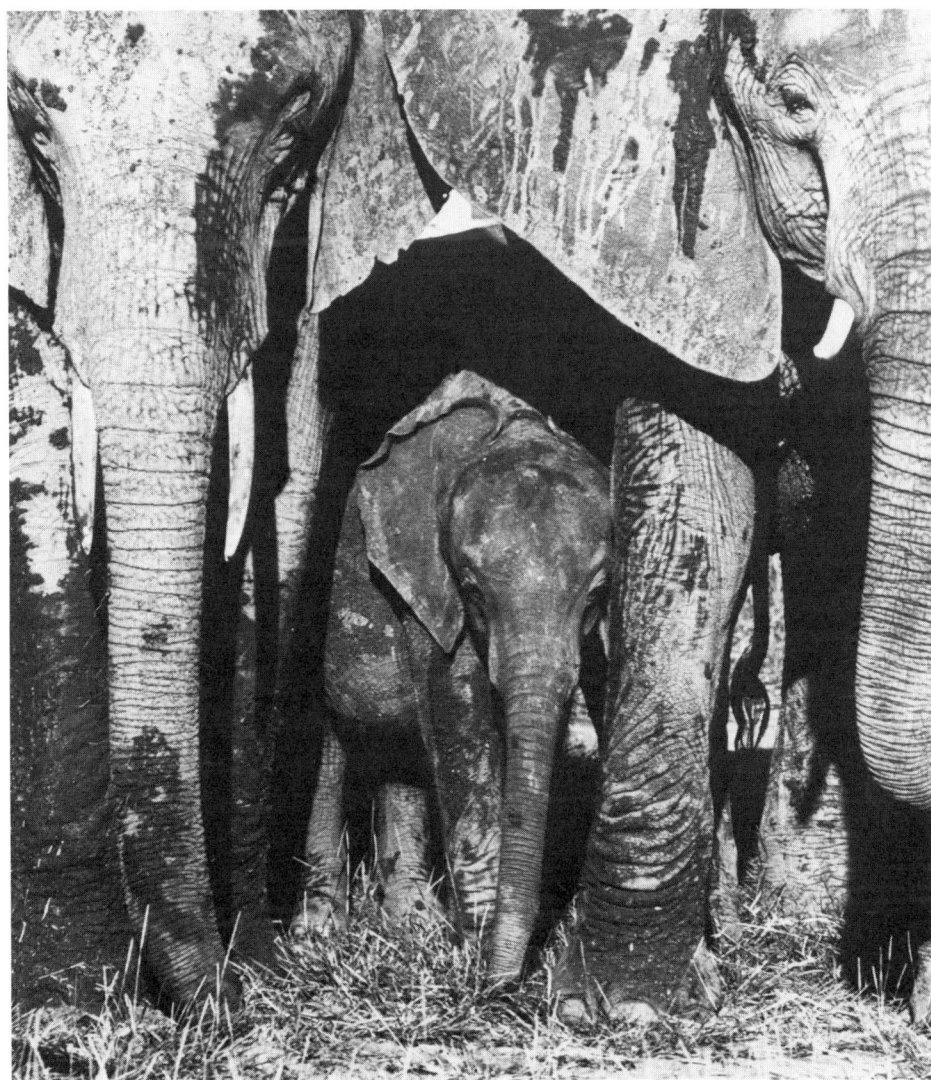

Right: A mother chimpanzee and her infant pause in a favorite activity—anting. Adults poke long grass blades into ant or termite hills. When the blades are pulled out, they are coated with clinging insects. Chimpanzees suck these blades like lollipops. *Above:* Here a baboon family engages in mutual grooming. Through grooming baboons keep their fur free of tangles and insects. But grooming is an expression of affection, regard, and friendship. Sociobiologists interpret activities like grooming as ways for a group to cement social ties.

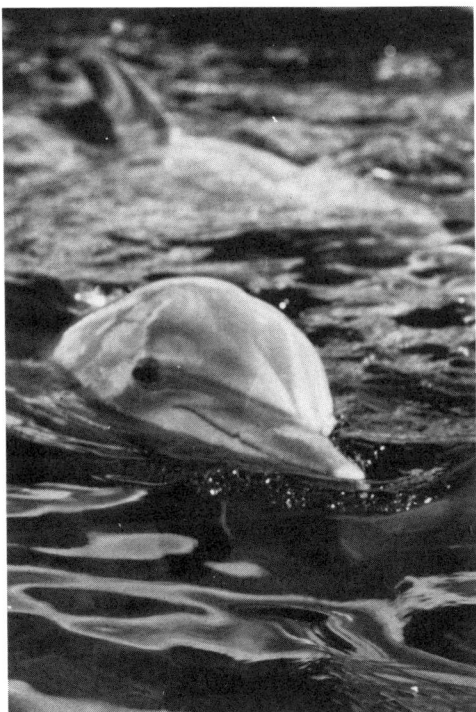

Dolphins are intelligent, generous mammals. But their altruism has caused problems for tuna fishermen. Dolphin herds often follow tuna schools and graze on fish in their wake. The dolphins are then caught in nets cast by tuna fishermen. If a dolphin becomes entangled, the others try to free it instead of fleeing before they are caught too. Many dolphins are drowned trying to rescue their companions. The United States Government is trying to prevent their destruction by designing new kinds of nets that the dolphins will easily leap over. Another way used to save dolphins is broadcasting the calls of killer whales, which frighten the herd away from the tuna as the fishermen approach.

Coyote social life seems to depend on food supplies. When food is scarce, coyotes band together. They also cooperate by bringing food, such as a rabbit, to pregnant or nursing females or to pups too young to hunt for their own meals.

male and his family that were captured and confined. At first, their behavior was considered typical. The male herded the females, which were very submissive.

Then the old male died. The young, cage-born males made only a halfhearted attempt to herd the females. Then they gave up the job altogether. What happened next could not have been predicted by sociobiology. An old female from the original group took over the role of the dead male. She herded the females and bullied them just as he had done. Yet she had once been a submissive female herself.

Finally, there are once-in-a-lifetime sightings that challenge previous assumptions about animals. Jane Goodall has watched the Gombe Stream chimpanzees for years. Except for struggles over dominance or occasional competition for choice food and tree nests, they live peaceably. Then, in 1979, she reported a most unexpected event. The troop split in two, and the first group fought and exterminated the second. If she had not seen this incident with her own eyes, she would not have believed chimpanzees capable of such violence.

A fourth criticism of sociobiologists' theories is that animals do have the ability to learn new behavior. A dramatic example comes from studies of monkeys on Koshima Island off Japan. Biologists scattered potatoes on the beach to lure monkeys into the open where they could be watched. The monkeys liked the potatoes but not the sand that clung to them.

One day a young monkey, a female the biologists had named Imo, took her potato to the ocean and washed it. Soon her playmates were imitating her. Then the adults imitated her. In time, young monkeys were taught from the start to wash their potatoes.

Later the biologists scattered rice instead of potatoes. It presented a new problem. The monkeys could not gather the rice without also getting handfuls of sand. Again Imo found a solution. She scattered the mixture on the ocean. The heavy grains of sand sank to the bottom while the rice floated on the top where she could gather it.

It is impossible to know if Imo was conscious of what she was doing or whether her discoveries were accidental. The important point is that she tried something new and realized that it worked. So did the others, and they soon imitated her. One animal had changed the behavior of the whole group.

A fifth criticism is that every animal action may not necessarily have a purpose. Some critics of sociobiology consider that assumption an outdated holdover from earlier times.

From Darwin's time on, many scientists studying evolution have assumed that natural selection makes animals better and better fitted for survival. Evolution was even described as "survival of the fittest." This idea fitted in with what people saw happening around them. New inventions like steam engines were improving daily life. It seemed logical to people that progress occurred steadily in nature too.

Ironically, Darwin was the one who tried to combat that idea. He dismissed the suggestion that change always means improvement. He deliberately referred to evolution as descent with "modification," not "improvement" or some other word that would suggest progress.

As a young man aboard the *Beagle*, he had seen evidence that change does not always serve a purpose. He had been collecting specimens of the giant tortoises that inhabit the Galápagos Islands. One evening he had dinner with the vice-governor of the Galápagos, who mentioned that each island had its own species of tortoise. Darwin was astonished—and embarrassed. He hadn't noticed the small differences in shell shape and pattern that distinguished each species and had mixed all his specimens up. He wrote in his diary, "I never dreamed that islands, about fifty or sixty miles apart, and most placed under a similar climate, and rising to nearly equal height, would have been differently tenanted."

46

Today a small but growing number of scientists is interested in the kinds of evolutionary changes that produced the different Galápagos tortoises. These scientists think that many important evolutionary changes serve no purpose. They occur when a small group of plants or animals becomes isolated and inbred. Certain genes for certain traits become concentrated in the small group. Which genes occur in the isolated population and thus have a chance to become predominant is a matter of accident.

Such accidents seem to account for the shell shapes and patterns of the Galápagos tortoises. Although all the species mingle in the water, they will only mate on the island where they themselves were born. At breeding time, they sort out and each species returns to the beaches of its particular island birthplace. This sorting out isolates them at mating time and allows genes for certain physical traits to become dominant in each species. In this way, each species remains distinct from the others, even though they all live in similar habitats and share a similar way of life.

These, then, are some of the objections to sociobiology. Such disagreements are common in science. Disagreement and debate are ways that new theories are tested and corrected. Most of the time nonscientists don't even realize any debate is taking place. That might have been true for sociobiology if sociobiologists had not turned their attention to human social life.

5

Human Behavior

The jump from elephants and chimpanzees to human beings is a huge one. But sociobiologists believe that their method of analyzing animal social life and cooperation can be applied to human behavior as well. They reason that since human beings share physical traits in common with other animals, they share behavior traits too. Sociobiologists compare human beings to baboons and chimpanzees, in particular.

Sociobiologists see human anger, jealousy, greed, and aggression as responses inherited from mammal ancestors. They see human generosity and loyalty as a form of altruism, advanced but nothing unique to the species.

In addition, sociobiologists think certain patterns of social life are genetically coded in human beings. One is a tendency to form social groups around kin. Another is an inherited difference in personality between males and females. A third is the tendency for males to mate with several different females.

Edward Wilson has speculated on how these tendencies might have evolved. Although very little is yet known about human origins, he has worked out his own reconstruction of what might have happened. He guesses that our early ancestors, tree dwellers in the forest, might have lived much like chimpanzees, our closest primate relatives. Their social life might have been a flexible one with mothers and their offspring forming close units, free to come and go as they pleased.

Then, Wilson suggests, later ancestors moved out of the forests onto savannahs and grasslands. There they faced dangers from predators and had difficulty finding food. Females and young needed protection. These hominids, or manlike creatures, began living in close bands. A baboonlike social group evolved.

There is no evidence that this description is what actually happened. In fact, scientists who specialize in the study of human origins have many different, often conflicting views on how human ancestors lived. Some scientists believe that early hominids would have been well able to protect themselves from danger. Even chimpanzees have been seen hurling rocks and sticks at a predator and driving it away. One scientist has even suggested that human body odor repelled predators and that hominids were therefore in little danger, even on the savannah.

Despite lack of evidence, Wilson and other sociobiologists are now trying to explain many features of human behavior as genetically coded traits. For example, they view homosexuality as a special form of kin selection. In this view, homosexuals provide an advantage to a family. They share the work of feeding and protecting the troop but produce few, if any, offspring. Thus, homosexuals provide extra food, attention, and protection to nieces, nephews, and cousins. Since human infants are totally dependent on adult care for years, this extra help might have meant the difference between life and death to early people. The homosexual's relatives would have had an increased chance to survive and carry on their benefactor's generous genes.

But there are characteristics of human behavior that sociobiologists cannot easily explain. Two of the most important are the role fathers play in human families and the enormous variation among human cultures.

In most species, the father has little to do with offspring. The bond that exists between a human father and his children is a special one in the animal world. Although male birds help feed and raise their young, only one other primate, the gibbon, forms such a family group.

Sociobiologists see this bond as a kind of compromise by the human male. His inherent tendency is to mate with several females and let each raise the young. But human infants are so helpless they need extra adult care. So the father must cooperate with the mother to raise his children. Sociobiologists believe the old instincts run deep, however, so that males still prefer to have several mates. On the other hand, they believe females are inherently content to be faithful to one mate. Further, they think females have become coy and flirtatious to attract and hold a male.

But this analysis leads sociobiologists to a second difficulty. In any group, there is a vast range of personality. Some women are not faithful, while some men are. Some women are strong-minded, while some men are submissive. Beyond this variety, human cultures differ greatly. In some, a man may have only one wife. In others, he may have many wives. There are even cultures where a woman may have many husbands.

Sociobiologists propose that this variation may be like racial variation in skin color or eye shape: genetically coded. Edward Wilson has even suggested that genetic coding makes people peaceful or aggressive, leaders or followers. In other words, nature accounts for most of the variation seen in human behavior and culture.

This belief has created a backlash against sociobiology. Wilson was even booed and hissed when he tried to speak at scientific meetings after *Sociobiology* was published. A group of scientists at Harvard, the Sociobiological Study Group of Science for the People, demanded a halt to all further research into sociobiology.

Scientists reacted strongly because many were alarmed at the political uses that might be made of sociobiology. A belief in inherited differences has long been used to justify oppression. Whites used it to justify enslaving blacks, who were said to be inferior. European settlers arriving in North and South America used it to justify depriving native Indians of their land. The Nazis used it to justify slaughter of Jews, Slavs, Greeks, and many other groups they considered genetically different.

Theories of inherited differences are dangerous for subtle reasons, too. They can provide an excuse to people who want to prevent social change. When women demanded the right to vote, men argued that females were mentally unfit for the strain of making decisions. Many people fear that sociobiology will provide a new excuse for sexist prejudice and will be used to oppose programs to help other disadvantaged groups. People may argue that nature, not lack of opportunity, accounts for their disadvantage.

Despite these objections, most scientists hesitate to ban sociobiology no matter how dangerous it seems. Such a ban, they argue, would destroy scientific freedom of inquiry. One sociobiologist spoke for many of her colleagues when she said, "I'm in the business of finding out the truth whether I like what comes out or not." She admitted that scientists can never be truly objective. A scientist with racist or sexist views will certainly interpret facts according to those prejudices. But she feels that debate and observation will expose such slanted interpretations.

Scientists also fear that such a ban would introduce censorship. It could prevent scientists from exploring potentially useful ideas. Scientists point out that Darwin's theory of evolution was considered dangerous in its time because it challenged the accepted Biblical account of the creation of the world. Censorship in Darwin's day would have hampered scientific attempts to understand how the natural world came into being.

So, instead of banning sociobiology, scientists prefer to study its evidence, debate its ideas, and allow independent judgment to emerge. Among the critics entering the debate are scientists who consider nurture important and who doubt that human beings can easily be compared to chimpanzees or baboons.

Such critics argue that sociobiology ignores the great differences between human beings and other animals, especially the human capacity for learning. Admittedly, other animals are capable of some learning. Elk, for example, learn migration routes from the herd's leader. And Imo showed that animals can create whole new customs like potato washing. But in all such cases the animals must learn by imitation. They must observe an action before they can perform it. Human beings, alone among animals, do not have to learn by imitation. They have language and writing with which to transmit their experience.

So great is the impact of learning, according to biologist George Gaylord Simpson, that human beings are "a fundamentally new sort of animal and one in which . . . a fundamentally new sort of evolution has appeared. The basis for this new sort of evolution is a new sort of heredity, the inheritance of learning."

Anthropologists have been studying this new sort of heredity for almost one hundred years. They have lived with different peoples, ethnic groups, tribes, and nationalities. They have recorded the many different ways of behavior in various cultures and how those ways are taught.

One of the world's foremost anthropologists was Dr. Margaret Mead. She provided interesting accounts of how nurture affects social behavior. From the 1920's to the 1940's, she lived with several different South Sea Island tribes. Some were cannibals and headhunters. Others were fishermen and traders. She recorded her observations in such books as *Growing Up in New Guinea* and *Coming of Age in Samoa*.

One group she lived with were the Manus, a tribe of fishermen whose children were very well behaved. She wrote: "In many societies, children's walking means more trouble for adults. Once able to walk, the children are a constant menace to property, breaking dishes, spilling the soup, tearing books and tangling the thread." But a Manus mother could spread out a heap of brightly colored beads without fear that her children would touch them. "Even the dogs," Mead wrote, "are so well trained that fish can be laid on the floor and left there for hours."

How did dogs and children gain such self-control? Adults constantly told them not to touch anything and punished them severely if they disobeyed. Mead herself found it "very tiresome to listen to the monotonous reiteration of some mother to her baby . . . 'That isn't yours. Put it down.'"

Nearby in Samoa she found a society with very easygoing standards concerning children's manners. Samoans think children are incapable of self-control until they are ten and develop *mafaufau*, common sense. A Samoan mother expects children to grab and break things, to throw temper tantrums, and to disrupt the peace of the household. Instead of disciplining their young children, Samoan parents make the ten- and eleven-year-olds responsible for keeping younger children away from adults.

As Mead described it, an infant "is fed when he is hungry, carried when he is tired, allowed to sleep when he wills. If he does wrong—cries . . . and disturbs the adults, defecates in the house, or has a temper tantrum—it is not he who is punished but the child-nurse whose duty it is to keep him out of such difficulties and to lug him out of earshot when he cries."

Could such cultural differences be genetically coded? Sociobiologists may think so, but Theodore Dobzhansky, one of the world's foremost geneticists and an expert on human genetics, did not. In his own book on evolution, he described why he disagreed with the sociobiologists. He thought the outstanding trait of our species is flexibility. In his opinion, what characterizes human beings is how *few* of our behavior traits are fixed, not how many, as sociobiologists think.

Dobzhansky is not the only geneticist to disagree with sociobiologists. Many point out that even physical traits are affected by nurture, despite genetic coding. Consider height, for example. Genes may code bone length and other factors that affect growth, but the end result also depends on diet and health. In Japan, children born since World War II are much taller than their parents. The reason is a change of diet. Before the war, most Japanese got little protein. Now their diet has improved, and they are growing taller.

Even unhappiness and emotional ills can affect physical growth. Doctors have now recognized that abused or unloved children can fail to develop properly. Doctors have no idea yet how unhappiness affects physical processes governing growth. They have named the problem the "failure to thrive" syndrome. It illustrates the complex and subtle relationship between nature and nurture.

Geneticists are learning that the effects of nurture begin even before a child is born. The United States Public Health Service is assembling evidence from many researchers that mothers who drink or smoke heavily, who take drugs, or who have severe emotional problems give birth to babies that are smaller than normal. These babies seem to have more physical and emotional problems as they grow up and often have trouble learning when they go to school.

Geneticists have come to think that genes can only code a pathway for potential development. Far from fixing personality and behavior in an absolute way, the use of the pathways is much determined by upbringing.

More information on behavior and social life comes from the experiences of the many races, ethnic groups, and nationalities that have immigrated to the United States. These immigrants have come from literally every corner of the globe. They represent the cultural variety that sociobiologists think might be genetically coded. Yet they all have the same complaint: Their children don't learn the old ways. They start behaving just like native-born Americans. Even Wilson admits this pattern is damaging evidence against his theory.

There is evidence that numerous sexual differences in behavior are also learned, not inherited. Each society has its own ideals of masculine and feminine. Margaret Mead wrote: "Sometimes one quality has been assigned to one sex, sometimes to another. . . . Some people think of women as too weak to work out of doors, others regard women as appropriate bearers of heavy burdens because 'their heads are stronger than men's.'"

She lived among the Tchambuli, a tribe that had only recently given up headhunting when she arrived in the South Seas. Tchambuli men were raised to be warriors. Yet they curled their hair or wore false curls if their own hair wasn't long enough for styling. They wore jewelry, spent the day gossiping, and did the family shopping. The women, who dressed simply and shaved their heads, carried on the community business of making and selling woven goods.

The Tchambuli assumed that men have no head for business. Little girls knew they would have to support their husbands when they grew up. Little boys knew they would not have to work for a living. Mead noted that Tchambuli girls of ten and eleven were more alert and inventive than the boys, while in the United States the reverse was true. She suggested that the expectations of each culture shaped the behavior of each sex from childhood on.

Even physical differences can be exaggerated by cultural beliefs. Men are generally bigger and stronger than women, but there is an overlap. Some women are stronger than many men. There is also an overlap in secondary sex characteristics like body hair, shape, and voice. In fact, in the 1960's, when boys let their hair grow and girls insisted on wearing jeans, adults complained they couldn't tell one sex from another. These adults were used to artificial clues like hairstyle and skirts to distinguish the sexes.

Perhaps all this evidence convinces you that arguments on the nurture side are conclusive. But matters are not that simple. Clearly human beings do retain some instincts from primate ancestors. Startle a baby, for instance, and it will throw its hands forward as if to grab something. This reaction is a very weakened form of the clutching instinct that enables an infant monkey or ape to cling to its mother's fur as she moves about. Such instincts are reminders of our common heritage with other animals.

For students of human behavior, the challenge is to strike a balance between a strict nature view and a strict nurture view. Back in the 1940's, biologist Julian Huxley described what he called the "nothing but" dilemma. He said that there are scientists who see people as "naked apes" and *nothing but* that. They emphasize human traits and actions shared with other animals. Then there are scientists who consider people as human beings and *nothing but* that. They emphasize the flexibility and learning capacity displayed in human development.

Undoubtedly, neither side will concede to the other in this debate. Perhaps the disagreement could only be settled if scientists raised some children from infancy without human contact to see what their behavior would be. But it is doubtful that anyone would perform such a cruel experiment. For without love and affection, the infants would not grow up to be normal adults.

6

The Final Word

Edward Wilson admits that sociobiology might be over-simplifying its comparisons among animals and between human beings and animals. Still, he thinks sociobiology is in an early stage of development and will work out its weaknesses in the next twenty to thirty years. Then, he predicts, it will lead to a true understanding of the evolution of animal social life and human nature.

But the history of theories about evolution casts doubt on that prediction. One reason is the immense variety of animal social life. There are over one million known animal species, and no two behave exactly alike. In some cases, scientists cannot even agree on a description of physical traits. So the prospect that they will agree on the descriptions of behavior needed by sociobiologists is unlikely.

Another reason is that the natural world is very complex. One perspective cannot explain all of evolution. Darwin had to pool his understanding of three subjects—botany, zoology, and geology—when he developed his theory of natural selection.

And even then he admitted that natural selection might not be the only mechanism for evolution.

Since then ecologists and geneticists have widened our understanding of the natural world. Now new fields are contributing still more insight. There are specialties like endocrinology—the study of hormones—or neurophysiology—the study of the pathways by which the brain directs the body's functions. Finally, fields like anthropology and psychology contribute information about human behavior.

As a result of this growing complexity, a single scientist in one field cannot possibly develop a theory that takes everything into account. Of course, that doesn't stop people from trying. As Huxley commented years ago, "Scientists often display a human failing: when they get hold of some new bit of truth, they are inclined to decide that it is the whole truth."

Another pitfall for scientists is the sheer volume of new information being revealed. In Darwin's time, the number of scientists was small. They could correspond personally with each other and exchange new information. Today the number of scientists is so large that even dozens of scientific journals and meetings aren't enough to transmit all that they are learning. The world of the scientist is like a kaleidoscope of changing information. Each scientist contributes a new bit to the shifting picture.

Given this situation, it is hard to predict what sociobiology will lead to in the end or how it will affect our understanding of behavior. Sociobiology has provided valuable insight into the role of kinship in many species, especially social insects, but it may not be useful in explaining cooperation and social life among all other animals. Its real value may be, in the words of one scientist, to create "profitable confusion," to challenge old views once more and force scientists to reexamine their assumptions.

In the end, the words Darwin wrote over one hundred years ago, in 1859, may always ring true. Apologizing for the gaps in his theory, he said, "No one ought to feel surprise at so much remaining as yet unexplained if he make due allowance for our profound ignorance in regard to the mutual relations of the many beings which live around us."

The final word? There is none.

Bibliography

* denotes books of interest to young readers

Beckoff, Marc and Wells, Michael C. "The Social Ecology of Coyotes." *Scientific American*, April 1980, pp. 130-148.

Darwin, Charles. *The Origin of the Species*. New York: The New American Library, Mentor Books, 1958.

*Darwin, Charles. *The Voyage of the Beagle*. New York: Dutton, Everyman's Library, 1959.

Dawkins, Richard. *The Selfish Gene*. New York: Oxford University Press, 1976.

Dobzhansky, Theodosius; Ayala, Francisco; Stebbins, G. Ledyard; and Valentine, James W. *Evolution*. San Francisco: W. H. Freeman and Company, 1977.

Futuyma, Douglas. *Evolutionary Biology*. Sunderland, Massachusetts: Sinauer Associates, Inc., 1979.

*Goodall, Jane. *In the Shadow of Man*. Boston: Houghton Mifflin Co., 1971.

*Goodall, Jane. "Warfare and Cannibalism Among Gombe's Chimpanzees." *National Geographic*, May 1979, pp. 592-620.

Greenberg, Les. "Genetic Component of Bee Odor in Kin Recognition." *Science*, 30 November 1979, pp. 1095-7.

Hamilton, W. D. "The Genetical Evolution of Social Behaviour, I." *Journal of Theoretical Biology*, July 1964, pp. 1-16.

Hamilton, W. D. "The Genetical Evolution of Social Behaviour, II." *Journal of Theoretical Biology*, July 1964, pp. 17-52.

Hamilton, W. D. "Geometry for the Selfish Herd." *Journal of Theoretical Biology*, May 1971, pp. 295-311.

Huxley, Julian S. *Evolution, The Modern Synthesis*. New York and London: Harper and Brothers Publishers, 1943.

Maynard Smith, J. "Group Selection and Kin Selection." *Nature*, 14 March 1964, pp. 1145-7.

*Mead, Margaret. *Coming of Age in Samoa*. New York: William Morrow and Company, Quill Paperback, 1961.

*Mead, Margaret. *Growing Up In New Guinea*. New York: The New American Library, Mentor Books, 1953.

*Mead, Margaret. *Male and Female*. New York: William Morrow and Company, 1967.

Ricklefs, Robert E. *Ecology*. Newton, Massachusetts: Chiron Press, Inc., 1973.

Rowell, Thelma. *The Social Behaviour of Monkeys*. Harmondsworth, Middlesex, England: Penguin Books, Ltd., 1972.

Schaller, George. *Golden Shadows, Flying Hooves*. New York: Alfred A. Knopf, Inc., 1967.

*Schaller, George, *Serengeti: A Kingdom of Predators*. New York: Alfred A. Knopf, Inc., 1972.

Schaller, George. *The Serengeti Lion*. Chicago and London: The University of Chicago Press, 1972.

Simpson, George Gaylord. *The Meaning of Evolution*. New Haven: Yale University Press, 1967.

Smith, Neal G. "The Advantage of Being Parasitized." *Nature*, 17 August 1968, pp. 690-694.

Tinbergen, Nikolaas. *Social Behavior in Animals*. New York: John Wiley and Sons, Inc., 1953.

Trivers, Robert L. and Hare, Hope. "Haplodiploidy and the Evolution of Social Insects." *Science*, 23 January 1976, pp. 249-263.

*Van Lawick, Hugo and Goodall, Jane. *Innocent Killers*. Boston: Houghton Mifflin, Co., 1971.

White, Tim D. "Evolutionary Implications of Pliocene Hominid Footprints." *Science*, 11 April 1980, pp. 175-176.

Wilson, Edward O. *The Insect Societies*. Cambridge: Harvard University Press, Belknap Press, 1971.

Wilson, Edward O. *On Human Nature*. Cambridge: Harvard University Press, 1978.

Wilson, Edward O. *Sociobiology, The New Synthesis*. Cambridge: Harvard University Press, Belknap Press, 1975.

Wynne-Edwards, V. C. *Animal Dispersion in Relation to Social Behaviour*. Edinburgh: Oliver and Boyd, Ltd., 1962.

Index

* indicates illustration

African wild dog, 24-26, 38*
baboons, 29-31, 35, 36, 42*, 45
Beckoff, Marc, 26
behavior: animal, 10, 14, 16, 21,
  23, 34-45; human, 11-12, 48-
  56. *See also* competition,
  cooperation, sex, social life
birds, 14, 34, 50; chestnut-
  headed oropendula 20-21;
  cowbird, 20-21; cuckoo, 20;
  turkey, 15-16
chimpanzees, 9, 10, 14, 31-32,
  42*, 45, 48-49, 52
competition, 14-15, 26, 34-35,
  45; dominance, 29-31, 35, 45;
  territory, 20, 24, 29, 36, 38.
  See also behavior, cooperation, sex, social life

cooperation, 9-10, 14-15, 19, 21,
  22, 25-31, 35-44*, 37*, 38*,
  39*, 40*, 41*, 42*, 43*, 58;
  caring for young, 19-21, 22-
  32, 35, 49, 53; protection,
  14, 15, 16, 22, 27, 29-31,
  34-35; providing food, 15-16,
  22-32, 36, 49. *See also*
  behavior, competition, sex,
  social life
coyotes, 36, 44*
culture, 11, 21, 50-56. *See also*
  nature versus nurture
Darwin, Charles, 13-16, 46, 51-
  52, 57-58
Dobzhansky, Theodore, 53-54
dolphins, 9, 29, 33, 43*
Douglas-Hamilton, Iain, 28

elephants, 9, 14, 27-29, 40*-41*, 48
elk, 52
Estes, Richard, 25, 26
evolution, 12, 13, 15, 46-47, 48-50, 51-52, 57-58; kin selection, 15, 20-21, 23, 25-27, 49; natural selection, 9, 13-15, 46. *See also* genetics
Franklin, William L., 24
genetics, 13-14, 16-19, 17*, 23, 27, 35, 47-50, 53-54; altruism, 14, 34; generous genes, 14, 34, 49. *See also* evolution, nature versus nurture
Goodall, Jane, 31, 32, 45
Greenberg, Les, 18-19
habitat, 10, 23-24, 31, 36
Hamilton, William, 14, 22-23
heredity. *See* genetics
human beings, 11, 48-51, 52-57. *See also* behavior, culture, nature versus nurture
Huxley, Julian, 56, 58
hyenas, 14
kinship, 16, 18, 28, 32, 58; family, 24, 25-32, 50. *See also* evolution
Koford, Carl, 24
lions, 14, 26-27, 33, 35, 39*
Maynard Smith, J., 14-15

Mead, Margaret, 52-53, 55
monkeys, 45-46
nature versus nurture, 11, 12, 52-54, 56. *See also* culture, genetics
prairie dogs, 14, 37*
pronghorn antelope, 14
Rowell, Thelma, 36
Schaller, George, 26, 27, 35
sex, 11, 15-16, 19-20, 23-32, 35, 36, 45, 48, 50, 55-56. *See also* behavior, competition, cooperation, social life
Simpson, George Gaylord, 52
Smith, Neal G., 20-21
social insects: ants, 9, 16-19; bees, 16-19, 17*, 21; wasps, 16-19, 21
social life, 9-11, 15, 21, 22-33, 35, 36, 37*, 38*, 39*, 40*, 41*, 42*, 43*, 44*, 47, 48, 52-54, 57. *See also* behavior, competition, cooperation, culture, sex
Tinbergen, Niko, 20
three-spined stickleback, 20
Trivers, Robert, 15
Van Lawick, Hugo, 25, 26
Vicuña, 23-24, 27, 29, 38*
Wells, Michael, 26
Wilson, Edward O., 21, 48-50, 55, 57